Wall Tie Failure

CAVITY WALL TIE FAILURE

The identification, treatment and repair
of buildings that have been affected
by the failure of metal wall ties

Malcolm Hollis
Professor of Surveying, University of Reading

1990

**Estates
Gazette**

THE ESTATES GAZETTE LIMITED
151 WARDOUR STREET, LONDON W1V 4BN

First published 1990
ISBN 0 7282 0162 3

© Malcolm Hollis/Leaf Coppin 1990

*This book was commissioned
and produced for the Estates Gazette Limited
by Leaf Coppin Publishing Ltd.*

Printed by Hobbs the Printers of Southampton

CONTENTS

Appendices

ILLUSTRATIONS

Introduction

A cavity wall consists of two leaves of brick or block with a continuous cavity in between, tied together across the cavity by means of a wall tie. The wall tie can be a galvanised steel, stainless steel or other metal or plastic device which is built into both leaves of the cavity wall to tie them together. This book will deal with the failures that may be caused by the deterioration of this tie between the two leaves which form the cavity wall and the repairs that are necessary. I will concentrate on the metal tie, the reasons for its failure, the identification of the failure and the various methods of repair when the tie is linking leaves of either brick, blockwork, stone or a mixture of these or other walling material.

The cavity wall is not a modern idea. The Egyptians, 4,500 years ago, formed walls by linking or tying together leaves of different wall material. Many of the pyramids were originally faced with a limestone cladding on top of the masonry construction, the cladding being held in place by non-ferrous ties.

Cavity walls were used in the Elizabethan and Stuart periods. For instance, the external facing in brick of a timber framed building required links to be formed to stabilise the construction. Georgian buildings were often formed by placing an external leaf of brickwork on the face of thicker brickwork. This external skin, usually erected in more expensive bricks by the best bricklayer, was tied to the remainder of the wall with headers running through the wall.

These early constructions resulted in a cavity being formed by default. The purpose was to mix different building materials either for the sake of appearance or to cut costs. In Victorian times, the benefit of the inner and outer part of a wall being separated by a cavity was recognised. These Victorian cavity walls were linked with cast iron, clay pipes, tiles or special cranked bricks, the bricks having been designed to reduce the transmission of water through the link. They were cranked to provide a fall from the inner leaf to the outer.

During the nineteenth century, metal ties were developed in a range of different shapes. Many were similar to the modern twist tie, and were forged in cast iron or unprotected steel. Some were dipped in bitumen to give them a protective coating. These early ties had weaknesses which were discovered with the passage of time. The brick ties or cranked brick links were vulnerable if movement took place in the wall because they were brittle and tended to crack. The cast iron ties performed well but are now approaching the end of their life. Wrought iron, sometimes with galvanising, has produced a lasting product which has performed better, but the inflexibility of the tie often damages the wall material, particularly if it is not very strong. The various types of tie used in buildings over the last hundred and fifty years are discussed below (see p. 18).

The modern form of cavity wall dates back to the end of the nineteenth century. It is probable that the earliest ones were located in buildings near the south coast, where there is a high risk of exposure to driving rain. The introduction of the cavity was intended to strengthen the resistance of the wall to the weather. The improved warmth that resulted was probably an unintended benefit, rather than a reason for adopting the process. In London the cavity wall was not adopted until the 1960s, but the rest of the country was quicker to introduce it.

Recognition of a Cavity Wall

Although a cavity wall is usually linked with the appearance of stretcher bond, determining whether the wall is a cavity wall is not always straightforward. The use of snapped headers in an external leaf has tended to confuse the observer when the construction of the wall is being determined. The bond can only be used as a guide to the construction and not as proof of the type of wall.

It is usually assumed that a cavity wall can be identified by the thickness of the wall, but that can range from 235 to 330 mm. The dimensions of a wall may vary because of the tolerances in the dimensions of the component parts, and it is usual to face the interior of the wall with a plaster coat, which may also vary in thickness. The width of the cavity can range from 40 to 70 mm and the inner leaf may be 75 mm, 100 mm or 150 mm block; the outer leaf may also vary in thickness and may be faced in render. Unfortunately, therefore, one cannot rely upon the measurement of the wall thickness as proof of its construction.

The measurement between the inner and outer faces of the wall can help to suggest the construction, and the measurement should be taken in all cases. A single brick wall is unlikely to measure, inclusive of plaster, in excess of 250 mm, whilst a cavity brick wall is unlikely to measure less than 290 mm inclusive of plaster.

The location of airbricks built into the external face of the wall should be noted. In some cases, the wall thickness can be measured through the vent, and occasionally the construction can be seen if there is no closing of the cavity above the air brick. The open perpend joint, located over a window or a door opening, is usually an indicator of cavity construction in a modern building. This weep hole that allows the discharge of water from the cavity gutter is recommended by Code of Practice 121, Part 1, clause 3.5.3. However, open joints over windows and doors are also left if put-log scaffolding was used and the openings were not filled when the scaffolding was taken down. Weep holes are usually regularly located, two to the head of each opening, whilst scaffolding support holes are more random, and not always located over openings. In the south-west of England cavity trays are often built in at first-floor level, and their presence at that position may help to identify the probable wall construction.

If the open joint to the perpend is below the damp proof course, it is probable that the building is timber-framed. That will mean that the external brick or block wall is facing the timber load bearing framework of the building. In such cases the ties will be more flexible and will be specially manufactured for use in such construction. The external leaf in such construction is not designed to carry any part of the load of the building.

The wall face and it components are often visible within the roof void. The cavity will be closed at eaves level, but is usually open to the upper edges of the gable, beneath the junction of the roof. It is generally possible to push your fingers into a gap to feel if there is a cavity. The presence of different walling materials on the inside and the outside of the wall is a good indicator of the probability of the wall having a cavity.

The inspection of the brickwork around door or window openings may help to identify the construction, as may the inspection of the unplastered surfaces often found under stairs or in meter cupboards. If the construction cannot be determined with confidence it will be necessary to expose part of the wall or use a borescope to examine the presence or otherwise of a cavity.

Why use a Cavity Wall?

The tie, linking together various parts or components of a wall, can provide a number of advantages. It enables different materials to be used in the construction, for example, a decorative material to face the less ornamental structure of the building. In that way the more expensive material can be used where it is visible and cheaper materials for the remainder.

Light materials can face a structure. Materials which could not support the building can be used as an external cladding. Materials which cannot support their own weight can be used on the face of a building and can be supported by the ties bedded into the structure of the property.

A wall can be created that is stronger than the sum of its parts. The linking together of two leaves of 112 mm brickwork by metal ties spanning a 50 mm cavity creates a wall that is stronger than a 225 mm wall would be. For instance, an 11" cavity wall can be built higher than a 9" wall. The requirements for lateral restraint will not be so great.

The cavity wall can be built over greater lengths than a 9" wall before the need for buttresses, provided there is provision for movement in the construction. The use of Portland cement mortars has increased the requirement for movements joints. Certain materials, such as lightweight blocks, require the provision of movement joints every three metres if minor fractures are to be avoided.

The cavity wall has better sound resistance and better thermal insulation than a solid 9" brick wall and is more resistant to water penetration than a solid wall, because there is a space between the two leaves. That internal gutter prevents water passing from one leaf to the other. Cavity filling, such as insulation, reduces that resistance and in certain circumstances has the effect of giving the wall a solid form so that water penetrates the interior through the fill in the cavity.

The cavity has advantages in strength, cost, performance and weather resistance. It is these advantages that were recognised, towards the end of the nineteenth century, and which led to this form of wall construction being adopted for most low-rise buildings.

The Performance of the Cavity Wall

Strength

The strength of the wall is gained from the tie between the two leaves. The outer leaf protects the interior of the building, provides the decoration and the weather resistance. The loads imposed on the internal leaf are distributed through the wall, as long as the wall ties are in good condition.

Any variation in the relationship between these two leaves will lead to the wall acting as two single leaves and not as a unit. The failure of the wall tie will affect the performance of the wall. Not only will the leaves act independently because they are no longer linked together, but also the corrosion of the metal of the tie may distort the wall and cause loads to be applied concentrically. The load may fall on the inner leaf, or at an angle to the inner leaf, resulting in an overturning moment, which can lead to the failure of the already weakened wall.

Loading

The cavity wall performs in a specific way. The loads of internal floors and roof are supported on the inner leaf of the wall, but because of the link between the two leaves of the cavity, it is the complete wall which carries the load. The weight on the inner leaf is redistributed on to each leaf, so that the load strikes the foundations around the centre line of the cavity wall.

The failure of the wall to act as a unit will lead to foundations being loaded concentrically, because the load is applied to half the wall. If the foundations were understrength, this may place too great a strain on them and lead to failure. If the foundations have deteriorated, or were undersized, this redistribution of the load applied to the foundations may lead to a more serious failure.

The ratio of the wall thickness to its height will have been reduced by more than half. This will mean that walls which have little or no lateral support will be likely to suffer distortion. Flank walls with the staircase running alongside are particularly vulnerable.

The Result of Tie Failure

Instability   The wall components may become loose and fall from the face of the building. Lightweight cladding on the face of a structure is more likely to fail when placed under stress by high winds or in the event of flexing, which may be caused by earth tremors.

Redistributed load   The failure of the metal tie reduces the strength of a cavity wall and may result in the collapse of a structure if the redistributed loading is beyond the capacity of either leaf. The tie failure will mean that each leaf will act independently, making them more vulnerable if either leaf has to take all the load.

Increase in height The corrosion of the wall tie will result in the metal expanding. It is the increase in the dimension of the metal which causes cracking to the wall. This has the effect of raising the height of the leaf of the wall, usually the external leaf.

Bulging If the wall is restrained at the top, the wall will be unable to expand vertically, and the pressure will result in the external leaf developing a bulge. If the bulge develops to 50 to 75 mm, there is a risk that a section of the brickwork will fail, and fall.

Damage to the roof If the wall is not restrained, the outer leaf of the wall will gain in height but the inner leaf will not. This raising of the height of the wall results in the outer leaf of the wall picking up the loading of the roof. The roof should be supported at the junction of the tie and the rafter in a pitched roof. The plate is located on the inner leaf. If the load is picked up on the outer leaf it can distort the roof frame. It also adds load to the outer leaf which is more likely to bulge or distort.

Gutter misalignment If the roof frame is lifted, it will result in the eaves being raised. This will in turn lift the gutter supports. Since the failure of ties is not uniform (it is common for the damage to be greater on the upper part of the wall, and damage on the south and west sides to be greater than on the north and east) the uneven movement, or movement to certain elevations and not others, will lead to gutter falls being changed. This can result in water entry and damage to the interior of the building.

Changed falls to a flat roof Where there is a flat roof supported on the outer leaf of a cavity wall, the failure of the ties, if it occurs in the wall below the roof, will raise the supports to the roof. This may rearrange the falls to the flat roof.

Reduced weather resistance The cracking to the external leaf of the wall will make the wall less resistant to the weather. This can lead to water penetration, particularly if the building is in an exposed location.

Summary

Wall tie failure can cause:

    instability
    weakening of the cavity wall

    eccentrically loaded foundations
    a weakened pitched roof frame
    increased risk of outer leaf bulging
    rearranged gutter falls
    altered fall to flat roofs

Why do Wall Ties Fail?

The corrosion of the metal of a galvanised tie is usually quoted as
the cause of wall tie failure. Yet there are numerous other factors.
Failures in ties are caused by both static and dynamic failings.
Static failings originate in a fault before or during building and
require no outside agency to assist the eventual outcome. These
are buildings which are doomed from the start, and where it is only
the length of time before the failure occurs that is in doubt. Dynamic
failures result from an interaction or change of circumstances,
more or less unforeseen.

Static Failures

Design   Design failure is one of the most common static failings.
The positioning of ties is also important. There are two documents
which impose a duty on the designer and the builder to comply with
the requirements set out therein.

    Part A of Schedule 1 of the Building Regulations is concerned
with the strength of wall ties, their stability and resistance to
deformation of the building and its parts. The recommendations for
construction are given within the regulations. The requirements for
wall ties are set out in Approved Document A1/2: Section 1 of
Document 1/2 and Part C, C23 deal with ties. New ties must comply
with British Standard (BS) 1243:1978 Specification for metal ties
for cavity wall construction, unless conditions of severe exposure
exist. If that is the case the tie must be austenitic stainless steel or
a suitable non-ferrous tie.

    The British Standard Code of Practice 121 includes directions
for the spacing of wall ties. According to the CP, ties should be
positioned at intervals not exceeding 900 mm horizontally, and not
more than 450 mm vertically, for walls with a cavity not exceeding
75 mm. Clause 3.12.2. requires spacing of ties to be not less than
2.5 per square metre. There must be additional ties at the sides of
all openings so that there is one for each 300 mm of height.

Tie location for 50-75 mm cavity walls where no leaf less than 90 mm thick. Not less than 2.5 ties/m²

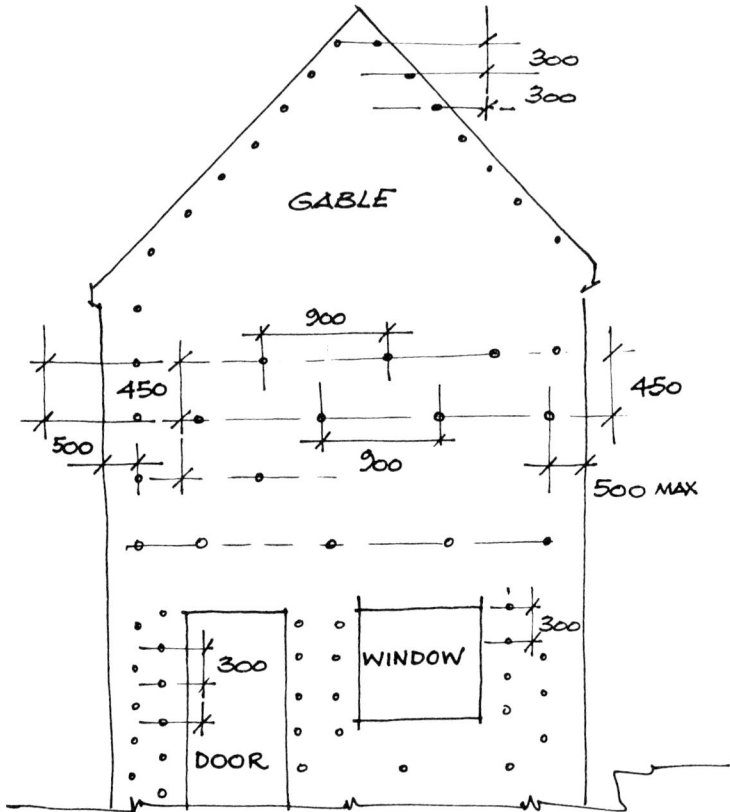

Workmanship   The failure to install ties of adequate quality, or to place them in the required position, will reduce the strength of the wall. If the bricklayer does not position the ties so that they are embedded equally in each leaf of the wall this will also lead to failure.

The design of the tie dictates its position. If it is a wire twist tie, the wire twist to the centre of the tie should be placed downwards. If it is an open loop tie, the dip in the link between each leaf should hang down. In both cases, the arrangement of the tie is to reduce the likelihood of water crossing the cavity along the tie.

There is debate as to whether, if ties are being inserted, a brick should be laid frog up or frog down. Bricks laid frog up will give a stronger wall, and the greater probability of the tie performing as intended.

The tie should be bedded into the cement. The cement has an alkali property which adds to the protection of the metal which is embedded in it, in much the same way as concrete protects its reinforcement. If the tie is lain on the top of the mortar joint there will be little protection other than the base of the brick on the next course. The strength of the tie fixing is not as good, and the metal tie will be less durable.

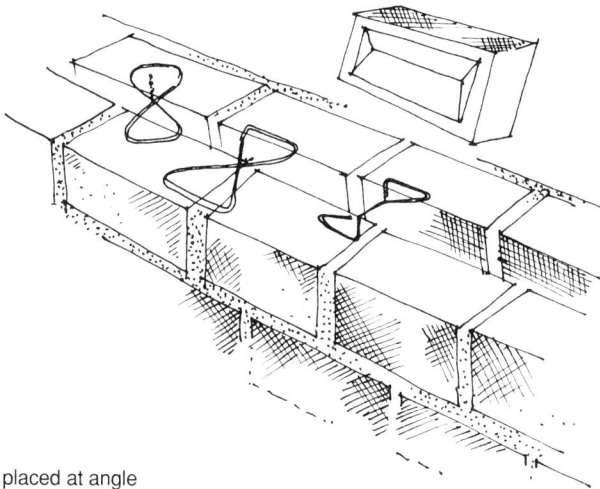

Tie placed at angle
Tie not placed centrally
Tie too short
Bricks laid frog down
Cavity width varies
Ties slope into inner wall

Chemical action   The chemical nature of the mortar is not always beneficial. If there are acidic, or potentially acidic, chemicals within the sand or the cement they will attack the galvanising on the tie. Black ash mortar has been known to act in this way: made from colliery waste the ash contains a lot of sulphur. When mixed with rain it oxidises to form a weak sulphuric acid which attacks the metal of the ties. Certain ties are more vulnerable than others.

Carbonation takes place in mortar, much as it does in concrete. The reaction of the mortar to the sulphur deposits, or to acid rain, converts the cement from an alkali to an acid. As it changes it no longer acts as a protector to the metal embedded in it.

Material failure   Wall ties for use in buildings are manufactured to BS 1243 (see above). Tests have shown that some ties have less than half the galvanising required by the British Standard. There is no easy method of checking the quality of ties delivered to site. The entry into the Common Market in 1993 will require much greater quality assurance in the manufacture of building components, and this may become less of a problem.

The requirements of BS 1243:1978, amended 1981, call for the following minimum quantities of galvanising:

Ties manufactured from	
zinc coated wire	940 g/m^2
plastics covered zinc coated wire	240 g/m^2
ties galvanised after manufacture	940 g/m^2

All ties are marked when in their bundle, but not every tie is marked that it is to BS 1243.

Dynamic Failures

These defects result from changes to the original circumstances of design and construction.

Fire   After a fire the tie will be more vulnerable to failure. The fire raises the temperature of the metal of the tie and damages the galvanising. The reduction of the zinc to the exterior of the tie will leave it, in part or totally, unprotected. The section of the tie in the cavity is the least protected and will suffer the greatest change during the fire and be the part most likely to fail through corrosion after the fire.

If a fire has occurred in a property, sample ties should be removed from different elevations for testing.

Alterations   Changes in the design of a building after completion require the work to be carried out in a way that will avoid the static failures enumerated.

The redistribution of load, resulting from the creation of new openings or the installation of beams or lintels, will cause some movement. That may distort a weakened wall. A failure to deal with junctions can lead to increased water penetration in the cavity, and hence to a greater risk of corrosion.

The addition of extensions can transform an exposed south facing wall into a sheltered wall which never receives any sunlight. This microclimatic change can lead to failure if, say, water penetration had been controlled by the drying effect of the south aspect. Condensation from kitchens or laundries, created within the cavity, may stay in the cavity rather than be moved through by the dynamic effect of circulating air aided by the temperature differential.

It is common to see new doors or windows cut into the original walls of a building. The alteration to the brickwork can be seen to have been no more than that required to allow the window to be installed. Clause 3.12.2. of Code of Practice 121 requires ties to be inserted at 300 mm vertical intervals around the window or door openings. These changes often do not make such provision, the many window companies selling replacement units, and carrying out this type of work on a package basis, not having operatives with the requisite knowledge of the requirements. The result is a weaker wall and an increased risk of failure.

Weather   If an exposed location exists at the time of construction, the tie has to be stainless steel. But exposed locations can be created through external agencies. Buildings which sheltered the property at the time of construction may be removed. New buildings may be erected which create a wind tunnel effect causing a greater degree of exposure to the wall.

The high winds of 1987 and 1990 have caused much damage to external wall leaves. The most vulnerable part of residential buildings was found to be the gable wall. The failures were often linked to inadequate workmanship at the time of construction. The high winds may only have accelerated the damage caused by the static failure that already existed, or they may have exposed the wall to pressures or suction that were beyond the design standards that were originally allowed for.

Walls which are weak in construction are likely to fail in severe winds. For instance, walls of different materials to each leaf require the coursing to be maintained. Where they differ, the ties cannot be inserted from one course on the inner leaf to the matching course in the outer leaf. It is in the upper parts of buildings where these minor faults are found. Ties folded into the cavity to avoid the problem at the time of building leave the wall weak.

Climate change If there were global warming, the higher temperatures would result in greater expansion of the components. Walls would have to have movement joints placed at reduced intervals. Existing buildings, which are frequently provided with inadequate facility for movement, would suffer. If there is differential movement between the inner and the outer leaf, there is a risk of the existing ties becoming distorted, and thus weakened.

Another effect of greater warmth might be the earlier failure of sealants, causing water penetration in the cavity, while movement in metal components might fracture the wall surface with the same result.

A reduction in the ozone layer could be harmful to plastics. Plastic ties may suffer if exposed to sunlight prior to being sealed up in the cavity wall.

Geological failures The earthquake that occurred in 1990 in Shropshire could be felt throughout most of England and Wales. The strong vibration must have damaged the bedding of ties in walls. The wall is a rigid structure not designed to cope with the extent of movement suffered that day. However, damage may go undetected.

Settlement in the ground, which causes foundation failure, may produce excess stress in the wall tie and lead to failure beyond the area of fracture and visible damage. In many city areas, the water table is rising and the variation in the volume of the ground will set up distortion unless action is taken. In the event that the rise in the water table is not countered, some ties in basement locations may become damp or, in extreme cases, be under water. The tie is not designed for that type of situation.

Vibration Buildings nowadays have to survive the vibration caused by modern living. Vibration can be caused by railways, heavy vehicles and, for buildings near an airport, aeroplanes taking off and landing. New vibration may be introduced by, for instance, the redirection of traffic closer to buildings, the construction of a new airport or the building of a new high-speed link to the channel tunnel. These new vibrations will change the circumstances for the wall. Different materials in the wall will vibrate at different rhythms. The smaller brick will not match the larger block. Stone of greater mass will vibrate in a different way. These unsynchronised movements will strain the tie. The flexing or distortion of the tie will weaken the galvanised coating and reduce the tie's performance.

Impact   Impact damage will occur if the wall is struck by another object, most probably a motor vehicle. More serious damage may be caused when a wall suffers minor impact at the time of construction. Impact from wind buffeting whilst the cement is green will reduce the quality of the fixings. Impact shortly after construction by other components may sever the fixing of a tie.

When there is impact damage, it is usual to repair the wall where damaged, re-using the existing ties, but the friction of the impact will have scraped the surface of the tie, removing some of the galvanising. This will reduce the life expectancy of the tie, which will no longer comply with the BS 1243 specification. This will increase the risk of failure. Furthermore, the exposure of the wall, between damage and repair, may have set up corrosion, which remains unchecked through the re-use of damaged materials.

The Corrosion of Metal Cavity Wall Ties

History of the Problem

When it was discovered that the cause of the failure of a wall in a Welsh farmhouse in the mid-1960s was corrosion of the metal ties, it was regarded as an isolated problem. It was nearly ten years before similar failures in other buildings were linked together by the Building Research Establishment.

By 1976 recommendations for the recognition and treatment of tie failure were set out in a book, *Common Defects in Buildings,* produced by the Property Services Agency (and published by HMSO). This book believed that the failure was restricted to buildings using metal ties and black ash mortar.

The recommendations for eliminating the problem were set out as follows:

a) The outer leaf can be removed and rebuilt. As it is likely that the remaining inner leaf will also have black ash mortar joints the new wall ties should be either stainless steel or plastic.
b) A cladding, such as tile hanging, can be applied to the wall. This will prevent rain penetration and may slow the rate of corrosion.
c) The wall can be opened up at each wall tie. If the latter is badly corroded it will need to be replaced, but if it is only slightly affected, so that when cleaned there is still sufficient metal to provide the functions of a wall tie, it could be painted with

bitumen paint and rebedded in a mortar not containing ash or clinker. The cost of this piece-meal work may be considerable.

d) The wall could be left as it is, providing that any water which passes through it does not spread into the inner leaf. However, the corrosion will continue and the ties eventually disintegrate, accompanied by wider cracks.

In 1979 two information papers were produced by the Building Research Establishment. One, IP 29/79, dealt with the method of replacement, using stainless steel rods, resin-grouted into the wall, whilst the other, IP 28/79: Corrosion of wall-ties: recognition, assessment and appropriate action, set out the problem and the method of its recognition. The Building Research Establishment had first drawn attention to wall tie failure the year before, but had deliberately avoided an alarmist tone. They knew that buildings constructed in certain periods would be liable to develop faults connected with wall ties. They had a moral duty to warn people but were concerned not to cause panic. This low key approach meant that for three or four years surveyors were not fully aware of the extent of the problem.

Meanwhile, in January 1982, the BRE produced a paper, Digest 257: Installation of wall ties in existing construction, which provided alternative methods of repairing failed cavities, and in March 1983, through their Defects Prevention Unit, they produced Defects Action Sheet DAS 21: External masonry cavity walls: wall tie replacement. It was this single sheet which was circulated as a press release by the BRE that attracted widespread media attention and finally alerted both the public and the surveying profession.

Recognition of the problem led to closer inspection of wall ties and their corrosion. The result was that the British Standard for galvanising metal on wall ties, BS 1243:1945 (which had been amended in 1964 to reduce the thickness of the galvanised coating required on metal ties), was further amended in 1978 and 1981 to require a greater thickness on the metal.

The 1945 standard had defined the allowable shapes of wall ties and the required thickness of the galvanising layer to be used on steel ties. The tie shapes that were specified included the vertical twist tie. This is a flat strip of mild steel with fish tail ends and a central twist to achieve a vertical section in the middle which is in the cavity once installed. It was the most common type used in prewar housing. The butterfly wire tie, creating two loops of galvanised steel wire joined in a spiral twist in the centre, had the benefit of being more flexible, and being less likely to damage softer and more fragile walling materials. In 1954 the double triangle tie was added. This is formed of stouter wire bent into two triangles with a straight wire section between. This design presented the smallest central section and thereby reduced the risk of mortar deposits building up on the tie in the cavity.

The butterfly wire, the vertical twist and the double triangle tie

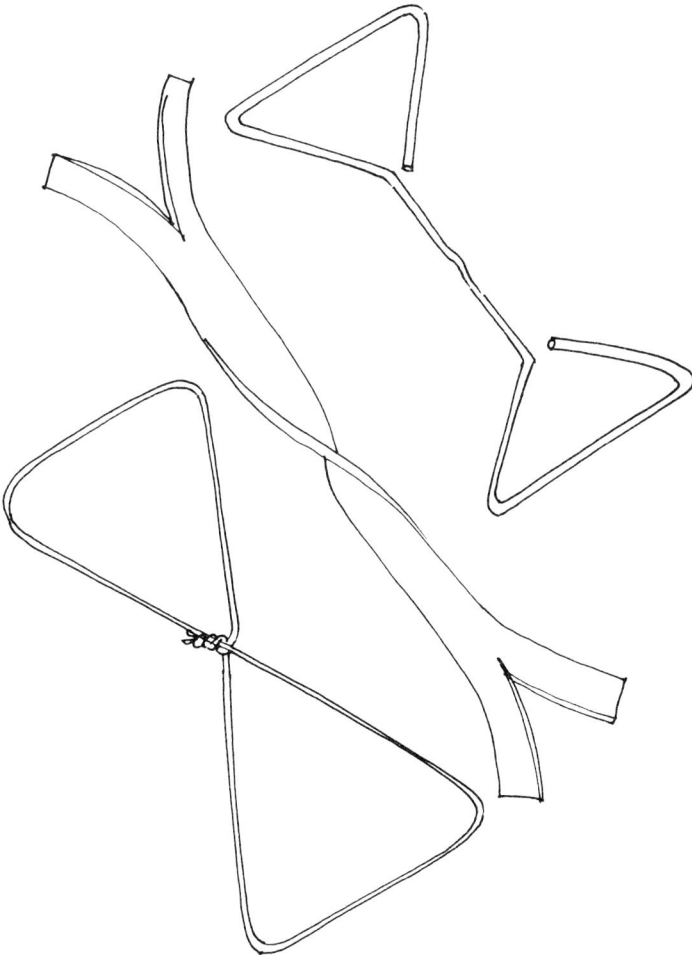

The ties could be made in two lengths, 150 mm and 200 mm. They could be in galvanised mild steel, copper or copper alloy, and, in a subsequent amendment, stainless steel.

The steel vertical twist tie was required to have a zinc coating of 540 g/m^2, whilst the wire tie was allowed to have a galvanised covering of half this amount. In the amendment in 1964 these standards were reduced to 380 g/m^2 for the twist tie and to 230 g/m^2 for wire ties. This reduction was made because of the satisfactory performance of these ties in service.

Following the discovery of failures in cavity ties, some of which had been produced to the lower standard, the standard was revised in 1981. The revised standard required a uniform covering of 940 g/m^2. This is an approximate increase of four times for the covering of a wire tie, and almost triple the 1964 standard for the twist tie. The standard now required is substantially higher than the standards that existed before the unfortunate revision in 1964.

These standards have extended the design life of wall ties. The pre-1964 tie had an anticipated life expectancy of the galvanised layer of forty years; this was reduced to twenty-five years after the 1964 amendment. The current standard will give a design life of sixty to seventy years.

The research of the BRE has shown that mild steel ties, protected with thin coatings of zinc galvanising or solvent-based paint, are not durable for the life expectancy of a masonry wall. Present experience now indicates that the following materials, to the specification in BS 1243, will have a reasonable life in normal exposure conditions:

Mild steel protected with a thick zinc galvanised coating of 940 g/m^2

Austenitic stainless steel grades 302, 304, 316 and similar

Copper, aluminium, bronze and phosphor bronze

Ties made before 1920 of wrought iron or cast iron located in sheltered areas away from the coast and not set in impure mortars

Injection moulded polypropylene for type 4 ties (adequate for two storey housing only, not suitable for larger structures because of their vulnerability in fire and creep susceptibility)

In locations of extreme exposure to driving rain, or wherethere is a significant chloride contamination, only grade 316 stainless steel, or plastic where appropriate, should be used

Conclusions

The risk of premature corrosion is greater for wire ties than for the flat twist metal tie when made before 1981, but buildings constructed between 1964 and 1981 will have a reduced life expectancy in their wall ties, with failure likely to occur within forty years of construction. However, the life expectancy of ties will be varied by the levels of exposure, or contaminants in the mortar, while the prediction of the condition of wall ties from an external examination of the building is likely to be imprecise.

Between 1880 and 1980 more than twenty million houses were constructed. The majority have cavity brickwork. There are no precise records of the number of buildings with cavity walls, but it is believed that there are over twelve million constructed with ties which have protection that is less than half the standard that has been required since 1981.

The adjustment to the recommendation for the spacing of the ties, as set out in the amendment to the BS 5628 in 1985, will not mean that ties inserted at different spacing in domestic buildings constructed before that date, with narrow cavities (of around 50 mm or less), will be seriously understrength.

Extent of the Problem

The full extent of the problem is not known. Failures are usually discovered when alterations are carried out to a building or when damage is caused, for instance, by impact or high winds. The corrosion of the heavier metal tie may be diagnosed because of the pattern of cracking on the exterior of the building, but the failure of the thinner wire tie cannot be discovered in this way. There is no national register of failure, and records maintained tend to be unreliable.

A number of companies have set up divisions which specialise in the replacement of defective ties. Their records of the areas where they are working provide a useful indication of the number of properties that are being repaired at a given moment. This information can be used to project the probability of damage being present in houses that were constructed contemporaneously, but the percentage risk of failure given below, for different ties, must be taken as a general guide only.

Cast iron   Used for wall ties in the nineteenth century. It has lasted well but is now found to be in a poor state. It is possible that tie failure will be found in seventy per cent of the buildings constructed in this way.

Wrought iron   Used between 1850 and 1920. This has performed well and is showing little corrosion. Some of these ties were coated in zinc. The presence of the zinc, or the deterioration in the coating, does not seem to have altered the ties' performance. It is probable that tie failure will be found in fifteen per cent of the buildings constructed in this way.

Mild steel   Came into use between 1920 and 1950. These ties were either unprotected, coated in bituminous paint or galvanised. The ties are now showing signs of failure, particularly in more exposed areas. It is probable that tie failure will be found in thirty-five per cent of the buildings so constructed.

Wire and flexible strip ties   Came into use between 1950 and 1981. These had less zinc protection because of variations in the standards set by the British Standard. The reduced surface area of the wire tie also meant that they had less protection. These ties are now vulnerable, and the risk of failure is considerable. It is probable that tie failure will be found in forty per cent of these buildings.

Ties since 1981   The standard for galvanising metal ties was revised in 1981. These ties are designed to have an effective life of sixty years. Stainless steel has been used to an increasing extent since this date. It is probable that tie failure will be found in three per cent of these buildings.

The location of the building will vary the risk of failure. Central city areas tend to be more sheltered and the ties to have performed better. Buildings in areas where there is mining are more vulnerable and have more instances of failure because of the use of waste products in building. Black ash mortar made from colliery waste has been used in Cardiff and other towns in south Wales, Liverpool, Sheffield, particularly in properties built between the wars, and in the Midlands. Whilst coal is the most common type of mining, damage has also been found in areas of tin mining. (Mundic blocks were built from tin mining waste and have been found to have a high level of caustic impurities, as well as being unstable).

Coastal buildings are more likely to have penetration of salt into the cavity and this will assist the agencies of metal corrosion.

The failure of ties in commercial buildings has followed a slightly different trend. It has been common to face buildings since the early 1960s. Masonry provided an attractive and relatively low cost facing to concrete buildings once the desire for a natural concrete material as external finish had evaporated. These buildings required the traditional tie to link the leaves of the cavity if walling was used on the frame. These wall panels were at much greater heights and were more exposed.

There was also a need for frame cramps to link the masonry to frames and solid walls and to restrain infill panels whilst making provision for differential movement. Failures have occurred, and these have led to a trend to use stainless steel. It is too early to evaluate the performance of these panels or the frame fixing. Greater exposure, larger areas of brickwork, and critical design tolerances have led to failures where all elements required in the design have not been met. The failure of the metal of the ties has also occurred.

Corrosion Chemistry

The failure of the metal tie is a chemical action. It is the result of the interaction of a number of elements. The definition of the main actions that take place is set out below:

Corrosion is the slow wearing away of solids, especially metals, by chemical attack. In the case of metals the mechanism is thought to be electrochemical.

Corrosion-fatigue is the phenomenon of the failure of metals subjected to repeated cycles of stress while exposed to corrosive attack.

Oxidation is the addition of oxygen to a compound, or a reaction which involves the loss of electrons from an atom.

Rust is a reddish brown oxide of iron (hydrated ferric oxide, $FE_2O_3.H_2O$) formed by the action of moisture and oxygen on metals. The added hydrogen and oxygen cause the expansion of the metal. The metal can increase in volume between seven and fourteen times.

The failure of metal wall ties through corrosion requires water and exposure to the air (see above, pp. 7-13, for the other causes of failure). Whilst cement is a natural protector of steel unless it has corrosive impurities, permeable mortars can be a threat to mild steel ties. Cement mortars are naturally alkaline and react with the steel to produce a protective layer of portlandite oxide on the surface of the metal. The chemical process of carbonation reduces the extent of the alkalinity, eventually converting the cement to a weak acid. Leaner, more permeable mixes speed up the carbonation process and also allow water to get to the tie.

Inorganic salts in the mortar, such as chlorides from marine sands, or in accelerators, such as calcium chloride, will increase the risk of corrosion taking place, while mortar containing black ash aggressively attacks steel.

In order to calculate the risk of chemical attack on the wall tie it is necessary to know:

the amount of protection on the steel tie
the chemical content of the mortar
the extent of carbonation in a cement mortar
the available moisture held against the tie
the permeability of the mortar.

Effect of Wall Tie Corrosion

Houses which have defective wall ties do not collapse overnight. The rusting of ties is a gradual process which slowly reduces the strength of the tie as corrosion progresses. Eventually the ties will fail. The wall then operates as two leaves, acting independently of each other. This will not lead to a collapse unless the wall is not in equilibrium.

The design of the building may have made it more vulnerable should the ties fail. Oversailing courses tied back to the inner leaf with wire ties will collapse if the ties fail. Walls that have distorted as the ties have expanded will fail when the ties go. The most vulnerable parts of the wall will be the smaller sections of unsupported brickwork, such as between two windows. The failure of wall ties will usually result in the upper part of the wall failing in the first instance.

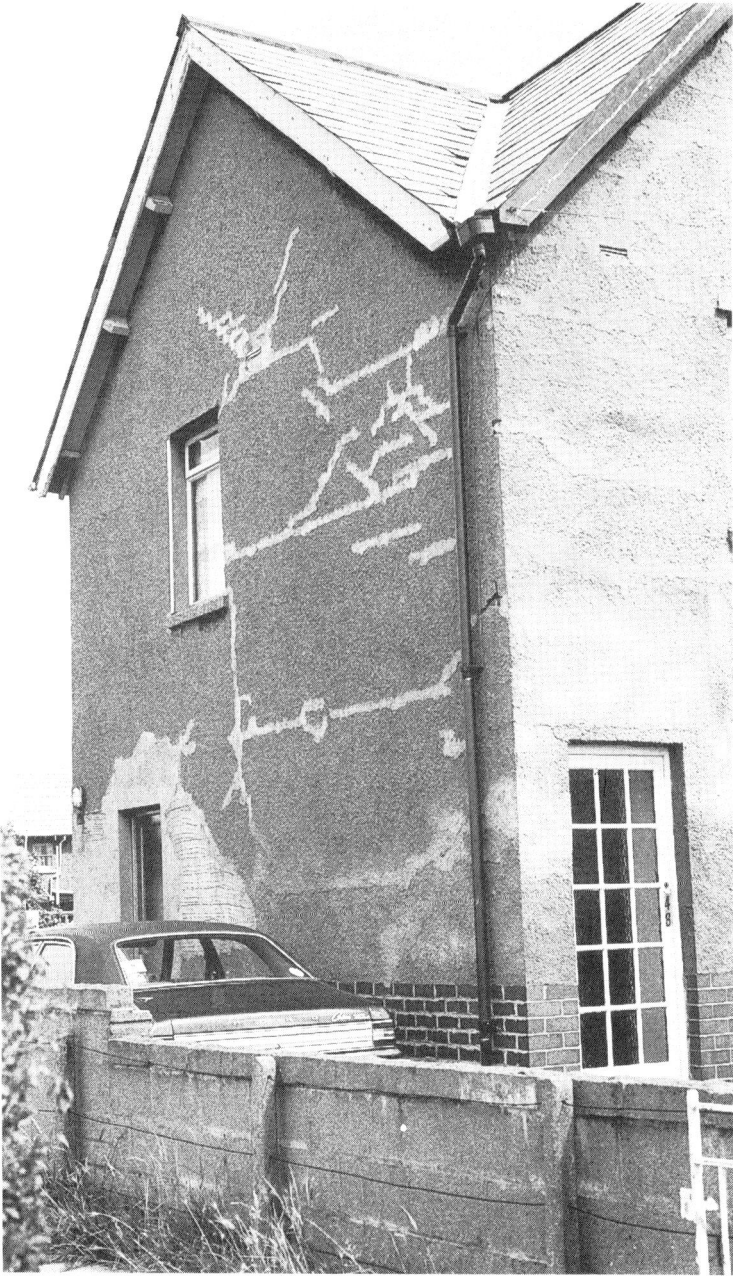

Detection of Tie Failure
Visible Damage

The expansion of the metal, caused by the corrosion, may produce defects that can be seen on the outside of the wall. This will only occur if the tie has a sufficient bulk of metal for its expansion to cause a significant fracture in the mortar joint. A twist metal tie will expand from its original dimension of about 5 mm to between 10 and 12 mm. This will result in an increase in the height of the external cavity leaf of about 75 mm (or 3"). The mortar joints every fourth, fifth or sixth course will have developed a crack of 5 mm. If the ties are spaced at greater distances, and there is a softer mortar, the disturbance will not be so obvious. In most buildings the failure will first be noticeable at the upper storey, but the total rise in the height of the wall will not exceed 40 mm. If the failure is allowed to develop, the increase in height will approach a probable maximum of 75 mm.

If the tie is a twist wire or double triangle, the bulk of metal will be much less. The extent of corrosion will not differ, but the force that it will exert may be insufficient to reveal any significant change on the external surface of the wall. The fact that this type of tie does not produce noticeable cracking in the mortar joints of the wall will result in the tie failure going undetected. However, this type of tie failure does not cause as much damage to the remainder of the building as that which can be caused by the corrosion of the twist metal tie.

Surface Inspection

The external face of the wall must be checked. This requires a close inspection of each brick joint, particularly to the upper part of the wall. This cannot be carried out by using binoculars. It will be necessary to raise a ladder and check each joint. The early indications may be no more than the splitting of the mortar joint from the underside of the brick to the course above. The failure may occur for a length of no more than 200 to 300 mm.

Some preliminary work can be carried out by inspecting the brickwork around upper windows. This must not be regarded as adequate to dismiss the risk of failure.

Location of Ties

In order to check the state of the tie it will be necessary to inspect a representative sample. The section of the tie which is positioned in a clear cavity can be checked visibly by using a borescope.

Borescopes   This is an optical instrument, placed in the cavity through a small pre-drilled hole. The borescope enables the tie to be viewed. The technique requires a considerable amount of practice in the interpretation of what one sees. The image is a function of the quality of the light source, the quality of the optics and the proximity of the hole and the tie being examined.

The condition of the section of the tie in the cavity is not a guide to the over-all condition of the tie. Failure in the mortar joint cannot be detected in this way.

The use of insulation in the form of cavity fill, both at the time of construction and installed at a later date has meant that many cavities are not clear, and this method of inspection is of no use. Because of the limitation of the borescope, the determination of the condition of a metal wall tie requires the removal of part of the wall to enable a physical inspection to be carried out.

The tie has to be located before the bricks around it can be removed for the inspection to take place. The tie can be located in a number of ways.

Metal detectors   Specialist metal detectors which will work on a field depth of about 100 mm have been developed, i.e. the machine is calibrated to concentrate on the location of metal in the thickness of the outer leaf of the wall. It will help to locate the position of the tie in a wall. It will not locate plastic or stainless steel ties.

Infra red thermography   Since there are variations in the temperature of the components of a wall, the checking of the temperature will help to locate those parts which are at lower temperatures than the rest. The wall tie conducts heat, so if the interior is warmer than the exterior, the tie should show up as a hot spot. The greater the temperature difference the easier to spot the ties. This is an expensive way of locating the ties in a wall, but will help to locate stainless steel ties, as well as the traditional mild steel tie. It will be unable to detect plastic ties.

Impulse radar   This is also an expensive method of locating ties and requires not only specialist equipment but also specialist knowledge in order to interpret the information obtained. The equipment will examine the surface of the wall, measuring variations in the materials within the depth of the wall. The readings are computerised at the time of the inspection and can be interpreted to show the position of ties. The trace will then have to be replotted on the wall to enable the tie locations to be marked.

Torch and mirror approach   An alternative to expensive equipment and specialist technical skills is to cut out a few bricks and make an inspection of the cavity with a strong torch and a mirror. The location of the course on which ties have been placed and the spacing can be estimated within one or two bricks of their actual location. However, there is greater cost in the opening up, and the wall will show the ravages of pock mark surgery.

Investigation

Once the ties have been located, a random sample of the ties should be removed. The type of tie must be determined. The range of ties falls into two categories, rigid and flexible.

Rigid ties   A rigid tie should only be inserted in strong walling material. The condition of the walling where the ties are located should be examined. If there is excessive damage, the wall may be failing because of an unsuitable tie, rather than because the tie itself has failed.

A rigid tie will be formed from a substantial section of metal, which has been flattened at each end to give it dimensions that will fit into a 10 mm mortar joint. The tie's performance will require a secure fixing into the wall at each end. Limited corrosion which reduces the grip of the tie will reduce its effectiveness.

To check the condition of the tie, the dimension of the metal's original thickness should be obtained. This can be done with a micrometer, such as is used for the checking or monitoring of crack sizes, the dimensions being taken from a section of a sound tie. The dimensional variation of the thickness of the tie will give information of the extent of corrosion, and the life expectancy of all the ties.

The approximate age of the tie should also be determined, so that the extent of surface galvanising can be checked with the standards that ruled at the time (see above, pp. 10 and 16, and Appendix 2) and its future performance assessed.

Flexible ties The flexible tie will be less stout, and is usually formed from wire (see above, p.15). The tie requires flexing to ensure that it is not damaging the walling into which it has been fixed. Corrosion can stiffen the tie before failure, and minor damage may have been caused to lightweight blockwork.

The body of the wire does not leave much tolerance to corrosion, and if any sign of rusting is found the tie should be replaced. The corrosion is an indication that the protective coating has failed, and failure of the tie will occur shortly. The location of the building and the level of its exposure will extend or reduce the remaining life of the tie.

The approximate age of the tie should be determined, so that the extent of surface galvanising can be checked with contemporary standards (see pp.10, 16, 50-1) and future performance assessed.

Mortar samples Samples of the mortar must be removed for analysis. (Addresses of laboratories that carry out this work are set out in Appendix 3). The presence and extent of any accelerators or caustic salts will help to determine the life expectancy of the existing ties, and to design the necessary repairs.

The mortar joints should be checked for variations in size. It was not uncommon for the bed in which ties were laid to be thicker than the bed of other courses. If there are fractures to each mortar joint of each brick course, this is more likely to be the result of sulphate attack within the brickwork. The pattern of any fracturing must be noted to see if it gives any indication that there are wall tie defects.

Material forming the wall The investigation must also identify the materials forming each leaf of the wall. The quality and performance of the inner leaf construction will limit and control the design of a replacement tie system, should one be required. For example, hollow blocks will present difficulties in the security of an internal fixing because of the limited amount of material to fix to; aerated blocks may prevent the use of rigid ties and limit the method of fixing because of their poor compressive strength. If the materials cannot be determined by surface inspection, then sections of the inner leaf will have to be removed for testing.

The size of the bricks of the external elevation should be noted. The range of brick dimensions may cause problems in the repair work. The move to metric dimensions is making it difficult to get matching sizes. The search for good second-hand bricks to match existing ones may be a time-consuming and expensive exercise.

Conclusion

The investigation that should be carried out to determine the extent of wall tie corrosion and to select the most suitable method of repair will include:

Identification of the tie, whether rigid or flexible
Noting the extent of damage to the performance of the tie
Noting the metal loss, if any, through corrosion
Noting physical damage to the wall materials caused by the tie
Testing mortar samples
Checking the thickness of the mortar beds
Identifying the materials of the wall.

Repair of Defective Ties

Once the failure of a wall tie has been established, it is necessary to decide how to carry out the essential repairs. That decision must take into account the client's needs and be cost-effective.

Does the Tie need Replacing?

Although a tie may have been found to be defective, this does not mean that it has to be replaced. It is possible that the wall is of adequate strength without the tie or that the extent of failure is not so great as to require immediate treatment. If the ties are flexible wire ties there is little likelihood of other damage being caused. In such circumstances, the inspection may give the client or house-holder the assurance that the repair can wait. If the building has a limited life, because it is due to be demolished, the repair may never be carried out.

If, however, there is any risk of partial collapse, it would be reckless to deal with the matter in this way. To establish the risk of wall collapse, the equilibrium of the wall must be calculated on the basis that the ties have failed. If that shows that a collapse could take place, the repairs must be put in hand.

Purpose of Replacement Tie

The aim of a replacement tie is to deal with one or more of the following circumstances:

to stabilise masonry cavity walls damaged by expansive corrosion of vertical twist or other type of steel wall ties;

to reinstate cavity walls rendered unsafe or unstable through loss of wire ties by corrosion;

to increase the number of ties in walls built with insufficient ties connecting the leaves;

to tie back existing cladding walls to concrete, steel or timber frame structures;

to stabilise faced or collar jointed walls in cases where the outer layer is becoming detached from the backing;

to tie new walls or bulging walls back to existing cross walls;

to tie back walls on either side of the cuts made for insertion of openings or movement joints.

Does the Tie need Removing?

The original tie may have corroded but not disintegrated. If it is a rigid tie, there is a further risk that continuing corrosion will cause damage to the remainder of the building. The lifting of the external leaf of the wall has various adverse effects (see above, p.6).

The alternative treatments for the existing tie are as follows:

Fold the tie down into the cavity The tie is left in the inner leaf of the cavity, and may corrode in the future.

Crop the tie in the cavity Remove the section in the outer leaf and dispose of it, leaving the section bedded in the inner leaf. This may eventually corrode, and may result in distortion of the inner leaf of the wall in the future.

Extract the tie Either pull it out of the wall or remove the bricks around its location, and lift it out. The quality of the walling will make this more or less difficult.

Isolate the tie Clan Contracting has produced a system, covered by patent, which involves the isolation of the tie in the external leaf of the cavity, so that any future corrosion will not cause further damage to the outer leaf of the wall.

The availability of these alternatives may be restricted by the nature of the particular building. For instance, it is not uncommon to find that the brittle nature of the internal leaf of the cavity has resulted in a rigid tie causing internal shear cracks in the inner leaf. It is not the tie which is the defect, but the relationship that it has with incompatible wall materials. In such a case, isolation may be required on the inner leaf, with a flexible tie replacement as the required repair.

Alternative Methods of Repair

Various methods of repairing damaged walls have been developed. The basic engineering requirements of the wall, and its replacement tie, are set out below. It should be borne in mind that the number of ties used can be varied to meet differing load requirements.

The selection of the replacement tie should be made on the same performance requirement basis as for the selection and installation of a wall tie in a new building (see Appendix 2). If one of the essential criteria is not met, the reason for the shortcoming should be examined to see if it will reduce the performance of the wall in the future.

In order to choose a tie it is essential that the nature of the building, its future and the intentions of the owner have been established. For instance, the life expectancy of the building and the owner's ability to pay for the repairs will have an effect upon the repair method that is selected.

The selection can be made on a number of judgements that might include:

Cost   The repairs possible for the money available.

Performance   The tie must make the wall capable of supporting a specific load or thrust.

Buildability   How long will the repair take? If the repair will cause disruption to a business, this may be the main consideration.

Appearance   What will the outside of the building look like after the work has been done?

Cost benefit   What is the future for the building, how long will the repair last?

Once the terms of reference for the repair have been established, it is necessary to select the best system of repair. The replacement tie must have the ability to:

transmit tension forces without excessive deformation;
transmit compression forces without excessive deformation;
transmit shear forces (particularly in timber-framed properties);
allow for vertical differential of movement of the two leaves (particularly in timber-framed properties);
allow for horizontal movement of the two leaves;
perform satisfactorily during a fire.

If the system has an Agrément Board approval it will have been tested and will meet these considerations. However, the advent of the single European market will see new products introduced into the United Kingdom. They may have been tested within their own country and meet their standards but while there is no uniform standard for such products, they will need to be carefully examined.
A wall tie should also be able to:

resist corrosion or other forms of degradation;
resist the passage of water from one leaf to the other;
offer as small a horizontal area as possible in the centre section to minimise the capture of mortar;
provide some flexibility to allow for inaccuracies in coursing;
accept the addition of a device for holding back insulation bats in some cases;
be rapidly and economically produced with efficient use of materials.

If the selection of a tie is made on cost alone, there needs to be a clear statement of the design restrictions that have been imposed, lest there be any doubt in the mind of the future purchaser or occupier of the property.

Limitations on Selection

If the cavity has been filled with insulation, the replacement of wall ties will injure this fill. In some cases this will cause no problem, but if the wall has been filled with a rigid polyurethane foam, the cracks caused by the tie installation may be ideal conduits for the transference of water from the outer leaf to the inner face of the wall.

Where the cavity has been filled with rigid insulation boards, the ties will have to secure the panels in the cavity. The presence of these panels will restrict the location, removal, cropping, folding down and even the isolation of the existing ties.

The presence of insulation in the cavity may also cause problems with the performance of the drip design in the replacement tie. If the insulation can prevent water running off, it may have the same effect as the tie being incorrectly located, with the drip against the inner face of the wall: water can enter the property and cause damage.

The nature of the bricks in the outer leaf of the wall must be established. The presence of large frogs in the bricks may mean that the point of drilling for the new ties needs to be carefully monitored in order to avoid forming a cavity which will not provide a secure fixing for the replacement tie. Brittle or friable bricks will prevent the use of ties that are fixed by the expansion of metal cramps on the end of the bolt, as these will fracture the bricks when the bolts are tightened up.

A similar examination of the material used in the inner leaf has to be undertaken, to see if that will impose any limitations.

If the outer leaf of the wall is seriously distorted or if the inner leaf has been disturbed there may be no possibility of repairing the wall by inserting new ties. In such cases partial rebuilding may be the only solution.

The size of the mortar joint may influence the decision to retain or remove the existing tie, even a wire twist tie. If the mortar joint is under 10 mm, the wire tie may expand and cause fracturing in the joint, even though in most cases the expansion of the metal may be insufficient to cause damage in a normal size joint.

Conclusion

The following is a brief check-list of the information required for the diagnosis and repair of wall tie failure:

    the inner leaf material
    the inner leaf construction
    the inner leaf condition
       any break up of the inner leaf
    the outer leaf material
    the outer leaf construction

the outer leaf condition
    extent of any bulging or cracks
the type of tie
the condition of the tie
the size of the cavity
the content of the cavity
any disturbance to the structure
any damage caused by tie failure to the building
    roof distortion
    gutter displacement
the exposure of building

Requirements of Replacement Ties

The tie must have an adequate fixing to the inner and the outer leaves of the wall   Certain replacement ties do not permit the testing of the fixing quality in both leaves of the cavity. The absence of security in either leaf will leave the repair vulnerable to further failure. The selection of the repair system must take account of these limitations.

The tie must be able to be removed if the fixing is inadequate Where the fixing can be tested, it is important that the tie can be removed if the fixing to either leaf does not perform to an adequate standard. If the design is irreversible and the fixing to the outer leaf, say, is poor, there is no way that the replacement tie can be removed.

The tie must suit the wall materials   The method of securing the tie to the wall must be compatible with the materials of the wall, and not cause damage or weaken the anchorage, or provide an anchorage of an unacceptably limited life expectancy.

Walls constructed with hollow blocks will require the selection of a tie specially designed for use in such materials. Walls built with low-strength materials may require glued ties because of the risk of failure under compressive stress if expanding anchorages are used.

A rigid tie can only be used to join strong materials. It is not suitable for use in block walls, or for joining brittle materials such as fletton bricks.

The tie must suit the construction   The tie required for a timber-framed building is very flexible and is fixed in a particular way, the design and its method of fixing being determined by the construction of the building. The design of lightweight concrete buildings also requires wall ties to perform in a particular way. The selection of a tie replacement must take into account any limitations imposed by the method of construction.

The fire resistance ability of the tie must meet the design requirements of the building   The determination of the requirement for fire resistance can be established by determining the reaction of the wall if the tie failed in a fire. If the failure of the tie or the failure and subsequent dropping of large sections of cladding, will lead to the immediate collapse of the wall, this will mean that the tie must have a particular fire performance. If the failure of the tie will not result in an immediate failure of the wall, such as may be the case in a domestic brick and block single or two-storey unit, the performance of the tie in the circumstances of fire may not be so critical.

Ties which have limited performance in the case of a fire are not only those in plastic, but also those fixed with neoprene gaskets. Tie systems with plastic components are limited to thirty-minute fire resistance unless specific fire test data stipulating a safe extended period is provided by the manufacturer.

The life expectancy of the repair must be compatible with the design requirement for the repaired building   The selection of a method of replacement tie should be linked to the desired life expectancy of the repaired building. The method selected could be varied if the life expectancy required is comparatively short, say, fifteen years. If the building is to meet a purchaser's requirements the repair should have a life expectancy of at least forty years.

The most durable of the materials used at the moment is Austenitic stainless steel. For the majority of applications, standard 18/8 grades such as 304 are satisfactory. The identification of stainless steel can be assisted by a magnet, which will not react to stainless steel, but it will not help determine the quality of stainless steel which has been used. Free machining grades of stainless steel contain phosphorus, which reduces the pitting resistance. For ties which may come in contact with salts, in exposed coastal areas, for instance, or where de-icing salts are used, grade 316 stainless steel should be used. This is highly resistant to corrosion. The complete tie should be made from one grade of stainless steel: if the quality is mixed in the components of the tie, there is a risk of galvanic corrosion.

When dissimilar metals are in contact with each other and in the presence of water or water vapour, which will act as an electrolyte, corrosion can occur. To prevent this, care should be taken to ensure that the metals that come into contact are compatible. If they are not, precautions must be taken to provide adequate insulation to prevent the metals coming into contact.

Table 1   Unsuitable metals to be in contact

Stainless steel	galvanised, mild and aluminium
Galvanised steel	stainless, brass, copper, nickel, lead and aluminium
Mild steel	stainless, copper, nickel and aluminium
Brass	galvanised
Copper	galvanised, mild and aluminium
Nickel	galvanised and mild
Lead	galvanised
Aluminium	stainless, galvanised, mild and copper

Table 2   Suitable metals to come into contact with each other

Stainless steel	nickel
Galvanised steel	none
Mild steel	none
Brass	copper and aluminium
Copper	nickel and brass
Nickel	stainless, copper and aluminium
Lead	none
Aluminium	brass and nickel

Table 3   Metals that are safe in contact only in dry conditions. They are unlikely to have these conditions in the United Kingdom

Stainless steel	brass, copper, and lead
Galvanised steel	mild
Mild steel	galvanised, brass and lead
Brass	stainless, mild, nickel, and lead
Copper	stainless, brass, and lead
Nickel	brass and lead
Lead	stainless, mild, brass, copper, nickel and aluminium
Aluminium	stainless and lead

Mild steel can only be used if it is adequately protected. One millimetre of a compatible and durable resin as an uninterrupted layer on the tie will suffice. The selection of such a tie must take into account the quality control of the protective layer, and the cost saving benefit of using such a tie must be compared with the risk, should the covering fail.

Replacement Wall Tie Systems

Replacement ties are usually selected for the amount of stiffness that is appropriate for the wall construction, the ability of the tie to fix into the particular wall and the cost involved. Ties vary from the flexibility of the wire twist tie to the rigidity of the vertical twist galvanised tie. But replacement ties tend to be much stiffer than the ties that are used in new construction. Expanding bolts are solidly anchored into each leaf and resin bonded ties are rigidly fixed to each leaf. The flexibility of a wire tie is only achieved by building in an equivalent tie as a replacement.

Each replacement method has limitations which will make the method unsuitable in certain circumstances. Those circumstances have been set out alongside each system or method referred to. Suppliers are listed in Appendix 4.

Rigid Replacement Tie Methods

Copper helical bar hammered into position   A hole of specific diameter is drilled through the inner and outer leaf of the wall and the twisted copper bar is driven into position using a purpose-made mandrel in a standard rotary-hammer drill.

The copper edges to the helical twist crush as they enter the hole and achieve a firm fixing to both leaves of the wall. The tie can be inserted into timber and low density brick or block.

Copper helical bar (manufacturer: Talbot Helifix Ltd.)

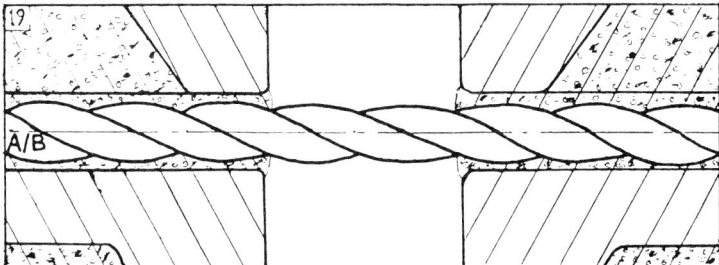

The quality of the fixing cannot be tested after installation for either leaf, nor can it be tested during installation. In practice, the application of hammering the tie into the wall tends to give the operator a feeling for the quality of the application through the resistance that is met during installation.

The installation of the tie may be unsuitable in walls that have become very weak, or where there is a risk of collapse caused by the vibration that will occur during installation. The hole size is critical to the quality of the installation.

The limitations are that it:

requires careful selection of walls to ensure suitability;
requires precision in drilling wall to ensure quality of hole size;
is difficult to monitor quality of work during application or after installation;
risks causing damage during application through vibration.

Grouted sleeved stainless steel tie (Manufacturer: WT Fixings)

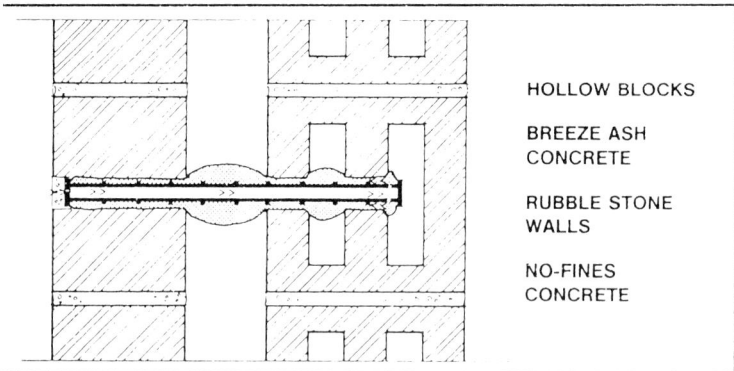

HOLLOW BLOCKS

BREEZE ASH CONCRETE

RUBBLE STONE WALLS

NO-FINES CONCRETE

Grouted sleeved stainless steel Manufacturers have developed varying techniques for the installation of ties that are resin bonded to each leaf. The systems deal with the design of the sleeve, and the method of getting the resin to the end of the tie in the inner leaf.

Each system is installed in a similar way: a hole is drilled through the external leaf and part way through the inner leaf of the wall. The stainless steel tie and sleeve are inserted into the pre-drilled hole, and resin is pumped into the hole through the sleeve surrounding the tie. The resin flows out of the far end of the tube and fills the hole in the inner leaf of the wall, gluing the tie and the sleeve into the leaf.

The tie is then glued into the outer leaf and the face of the wall is made good.

The quality of the fixing in each leaf of the cavity cannot always be checked during or after installation. The system is specialised, and requires experience in its installation if quality is to be achieved or maintained.

The limitations are that it:

requires careful selection of walls to ensure suitability;
is difficult to monitor quality of work during application or after
   installation.

Grouted non-sleeved stainless steel   A hole is drilled through the outer leaf of the wall and part way through the inner leaf. The hole in the inner leaf is filled with resin, applied through an extension nozzle, and a stainless steel bar is inserted into the hole and resin. The hole in the outer leaf is then resin filled and the external face of the hole pointed or faced to match the colour of the brick into which the tie has been inserted.

The fixing into the inner leaf can be tested before the external leaf fixing is completed. It is a simple system, which requires little specialised application. The selection of the resin may be determined by the requirement for fire performance.

There are variations of this system. For instance, where the tie is a hollow tube into which the resin is pushed, the tie may be glued

Grouted non-sleeved stainless steel tie (manufacturer: Pynford)

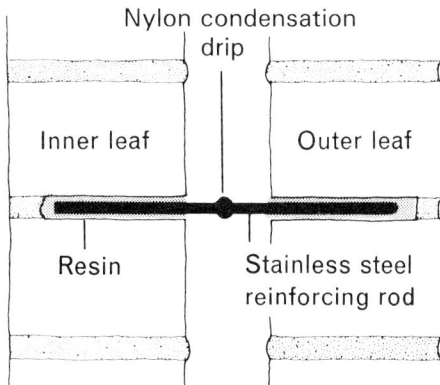

Nylon condensation drip

Inner leaf   Outer leaf

Resin   Stainless steel reinforcing rod

or bonded to the inner and outer leaf at the same time. In such a case it is not possible to test the fixing to either leaf at the time of installation, or afterwards.

The limitations are that it requires careful selection of walls to ensure suitability.

High density polyurethane foam   A heavy duty polyurethane foam is injected by specialist installers into the cavity with the intention that it should glue together the two leaves of the wall. The quality of the finished job depends upon the condition of the wall cavity before installation. The surface should be clean and dry, and, if that is to occur, the original ties must be removed before the foam is installed. The risk of water penetration may be high in areas of high exposure and must be considered before selection of this method of repair. The wall may have to be tied to prevent it spreading during the pressure application of the foam.

This method of repair was pioneered in south Wales. Early installations failed. The main cause of failure was the dirty surface of the inner leaves of the wall preventing adequate adhesion of the foam, or damp or wet external leaves having a similar effect. Unsatisfactory results also occurred where water penetration took place to the inside of the repaired property because of the application of foam in walls which were unsuitable through exposure ratings or the nature of the external leaf.

The limitations are that it:

requires careful selection of walls to ensure suitability;
requires a high standard of application;
is difficult to monitor quality of work during or after installation.

Stiff Replacement Tie Methods

Stainless steel screw-in anchors   A hole is drilled through the external leaf of the wall and part way through the inner leaf of the cavity. A stainless steel double expansion bolt is inserted into the hole and the expansion nut is tightened. This draws in the cone-shaped ends to the screw-threaded bolt, thus expanding the ends of the outer sleeve, and securing the bolt to the inner and outer walls in the same action.

The quality of the fixing depends upon the accuracy of the hole, the correct assessment of the durability of the walling materials and the avoidance of over-expansion of the bolt by over-tightening the expansion nut.

Stainless steel screw-in anchor tie (manufacturer: Harris & Edgar)

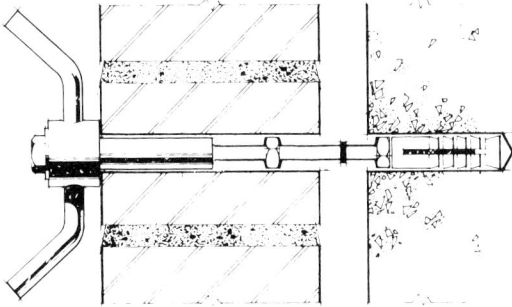

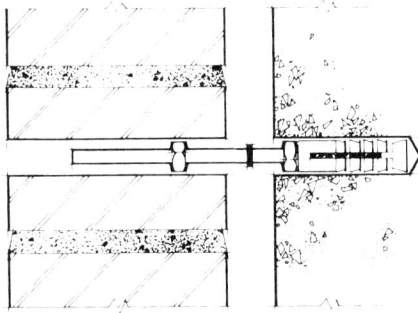

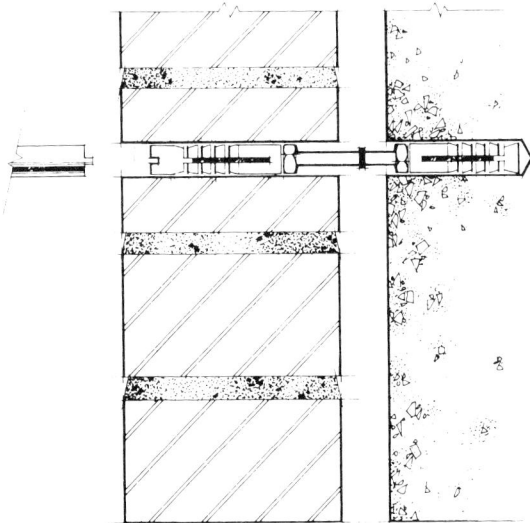

Stainless steel double expansion wall tie (PW – manufacturer: Hilti)

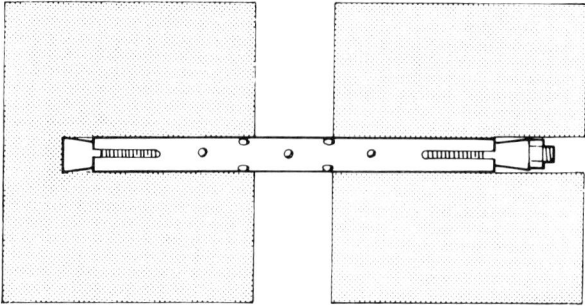

As above, with independently set mechanical anchors (PWS – manufacturer: Hilti)

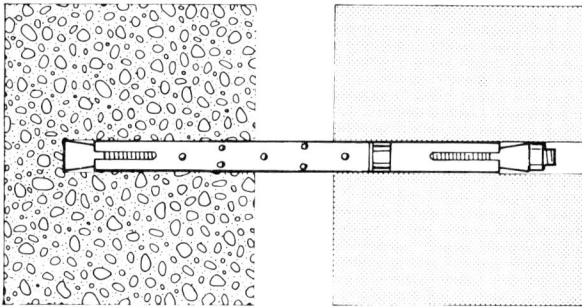

In this case, the fixing consists of a stainless steel body with aluminium bronze expanding shells at each end (Hemax 63 – manufacturer: Harris & Edgar)

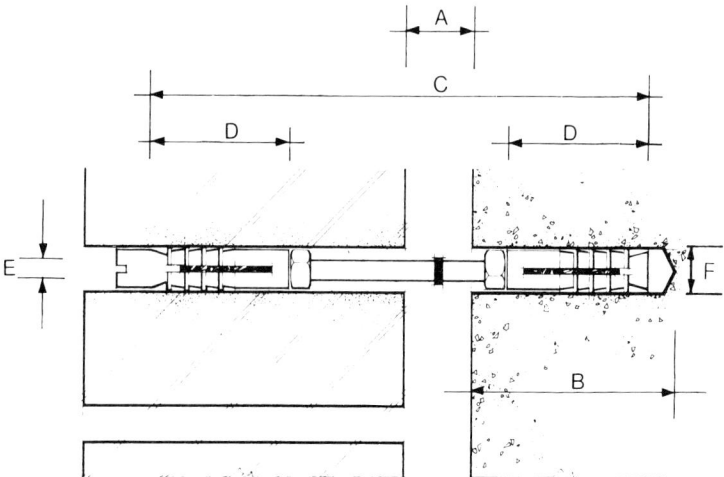

Variations of this bolt permit the tightening of the expanding section in the inner and the outer leaves independently. In this way, the bolt can be inserted from the inside of the building, avoiding any disturbance to the exterior of the property and the cost of scaffolding.

The limitations are that it:

requires careful selection of walls to ensure suitability;
may be impossible to test finish in external leaf;
may not be suitable for installation in hollow block inner leaf walls;
may be difficult to monitor quality of work after installation.

Combined bolt and resin ties   Combination ties have been developed for use where there is a low quality internal leaf, which would be unable to support the mechanical pressure of an expanding bolt.

A hole is drilled through the outer leaf and part way through the inner leaf of the wall. A bolt is inserted into grout applied to the hole formed in the inner leaf. The uneven surface of the shaft is secured to the inner leaf with a polyester grout. Once the grouting has cured, the bolt is attached to the outer leaf by tightening the expanding section in the outer leaf of the wall.

The finished tie is glued into the inner leaf and fixed into the outer leaf with an expanding bolt. The resistance of the resin to the passage of the tie can be an indication to the installer that there is enough resin in the inner leaf socket.

The resin will harden, but the curing time will vary, depending upon the temperature. At 41°F (5°C), resin will take 4 hours to achieve its full strength, whilst at 77°F (25°C) it will only take an hour.

The limitations are that:

it requires careful selection of walls to ensure suitability;
it may be impossible to test fixing in external leaf;
completion of installation cannot take place at same time as insertion of bolt;
fixing can be damaged by premature tightening of the bolt, before the full strength of the resin has been achieved;
it is not suitable for installation in hollow block inner leaf walls;
it may be difficult to monitor quality of work after installation.

Combined bolt and resin tie (Bondtie and Macbond – manufacturer: Red Head)

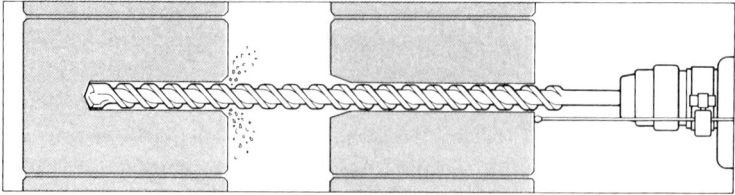

The depth gauge is set to allow the pre-determined drill bit to penetrate the inner leaf until the gauge meets the face of the outer wall

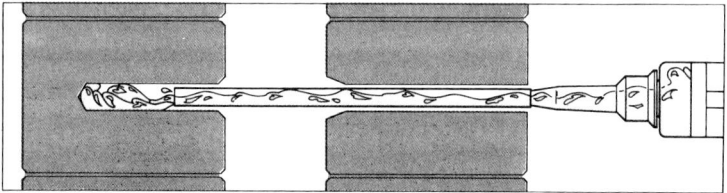

The delivery tube is passed to the back of the inner leaf. The trigger is squeezed to force out the resin, then the tube is slowly withdrawn until clear of the inner leaf. The pressure is released and the cartridge withdrawn

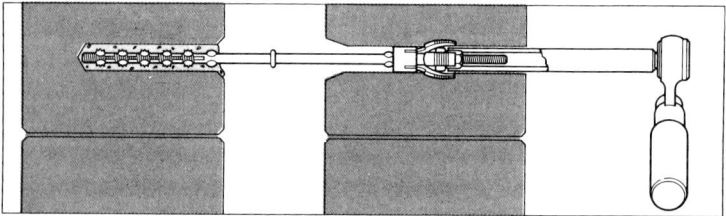

After initial set, the hex socket setting tool is fitted on to the torque spanner and the top expander unit is set by tightening the nut until the spanner breaks at the pre-set torque

Flexible Replacement Tie Methods

Stainless steel ties with plastic friction grips on both ends   A hole is drilled through the external leaf of the wall and part way through the inner leaf of the cavity. A stainless steel double expansion bolt is inserted into the hole and the expansion nut is tightened. This draws in the end to the screw threaded bolt, thus expanding the flexible gasket which forms part of the outer sleeve, and securing the bolt to the inner and outer walls.

The quality of the fixing depends upon the accuracy of the hole and the correct assessment of the durability of the walling materials. The flexible nature of the PVC sleeve spreads the load and reduces the risk of damage being caused to the walling material.

Variations of this bolt permit the tightening of the expanding section in the inner and the outer leaves independently. Thus, the bolt can be inserted from the inside of the building, avoiding any disturbance to the exterior of the property and the cost of scaffolding.

The limitations are that:

it requires careful selection of walls to ensure suitability;

it may be impossible to test finish in external leaf;

the material of the gasket may react with corrosive chemical that may be found in certain mortar joints. The bolts should be kept clear of the joint, but care should be taken over the assessment of the risk of the leaching of the acid content on to a nearby bolt, and the reduction of the life expectancy of the tie;

it may be difficult to monitor quality of work after installation.

Stainless steel tie with plastic friction grips on both ends (Fastie – manufacturer: Red Head)

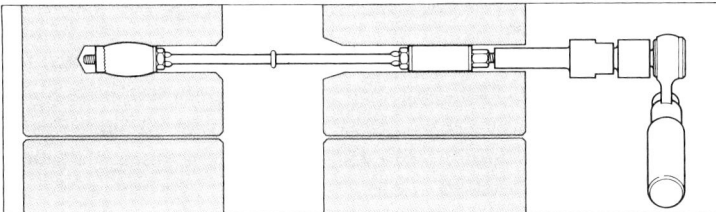

The torque wrench is turned to rotate the tie and expand the inner sleeve until the required torque is reached when the pre-set mechanism will break the torque spanner

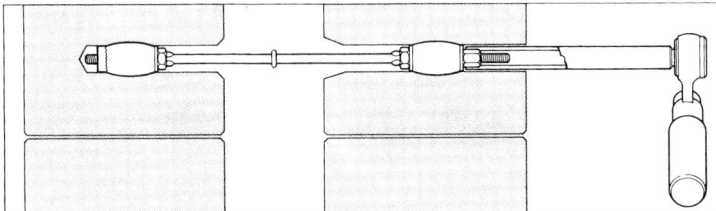

The hex socket setting tool is fitted on to the torque spanner and the top expander unit is set by tightening the nut until the spanner again breaks at the pre-set torque

An alternative stainless steel tie with plastic friction grips on both ends (Mactie – manufacturer: Red Head)

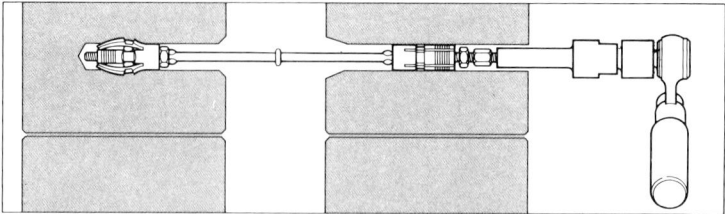

The torque wrench is turned to rotate the tie and expand the inner sleeve until the required torque is reached when the pre-set mechanism will break the torque spanner

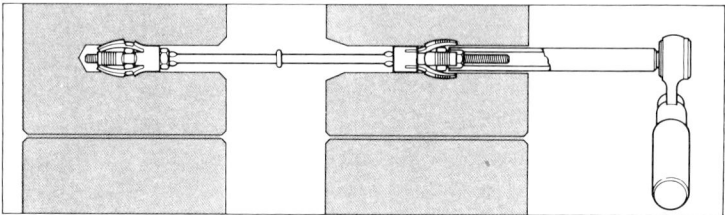

The hex socket setting tool is fitted on to the torque spanner and the top expander unit is set by tightening the nut until the spanner again breaks at the pre-set torque

Stainless steel grouted A brick or bricks are removed from the outer leaf of the wall and the inner leaf of the cavity is drilled. Resin is inserted into the pre-drilled hole in the inner leaf before the tie is bedded in. The tie is folded or adjusted so that the triangle sits in the re-bedded mortar bed in the external leaf. The brick or bricks are replaced in the external leaf of the wall and pointed up to match existing work.

The system uses a manufactured tie. There is a range of designs which includes ties that look like half the conventional double triangle tie and that have an elongated wire end which is either straight or cranked. It is this wire end that is bedded into the resin in the drilled socket in the inner leaf, while the conventional end is built into a mortar joint. The quality of the fixing in the inner leaf can be tested before the bricks are replaced in the external leaf.

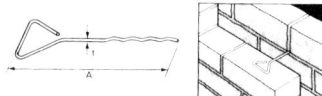

Stainless steel grouted tie (Perfix wavy tail – manufacturer: Hilti)

If the inner leaf is built with hollow or perforated blocks the fixing of the tie into the material will be difficult, and the quality of the fixing may be unsatisfactory. Inorganic grout should not be used if there is any requirement for a fire performance in excess of thirty minutes, or if the tie is to be used as a fixing into timber frame.

The limitations are that:

it requires careful selection of walls to ensure suitability;
it requires removal of bricks in the external leaf, and the finish shows signs of work to the external leaf;
it is impossible to test the finish in the external leaf;
the selection of the grout or adhesive in the inner leaf may be determined by fire or material considerations;
it is not suitable for installation in hollow block inner leaf walls;
it is difficult to monitor quality of work after installation.

The Role of the Surveyor

The duty of the surveyor carrying out an inspection of a property is to advise upon the liability that the property represents. To assess the condition of a building, the surveyor agrees the extent of the inspection and then investigates those parts of the property in the manner agreed. Wall tie failure is not always easy to identify. There may be no external indication of the problem although failure may be imminent. If the ties are twist metal, the metal may not have adequate body to cause cracking despite the increase in volume owing to rusting.

What can the Surveyor do?

If there is no indication in the exterior of the building that could suggest tie failure, the surveyor must inform the purchaser or owner of the risk of tie corrosion, and the damage that will result, in the following circumstances:

1. If there are known failures in cavity ties in the town, or immediate area in which the building is situated; the warning must clearly indicate the level of risk that can be attached to such a building.
2. If the building has in the wall construction corrosive materials that could be assumed to come into contact with metal ties; a clear warning of the risk that is present must be expressed if, for instance, the building is constructed with black ash mortar.

If the building shows signs of failures that could be an indication of wall tie corrosion, the surveyor will express a clear warning of the likelihood of tie corrosion. It should contain a number of elements:

The serious nature of the problem It is essential that anyone reading the warning is aware that the failure of wall ties is expensive to rectify, disruptive and damaging to the building, and will prevent the building being sold until the damage has been repaired, and the ties replaced.

The level of cost of repairs An indication that the repair of the wall ties may cost around £5,000 will give guidance to an owner or prospective purchaser of the type of problem that he faces. The knowledge that repairs may include damage to gardens and to internal and external decorations will give a measure of the disruption that will go with the elimination of the problem.

The level of certainty of the presence of the problem Surveys are not pathological inspections involving the opening up of all parts of a building to locate all defects and to determine the full extent of all problems that may exist. The diagnosis is often only sufficient for the surveyor to believe that there is a risk of a problem being present. The surveyor needs to inform his client of the level of inspection and the degree of certainty that he has over the presence of tie failure.

The actions available to identify the full extent of the problem The surveyor needs to advise the client of the actions that can be taken to determine the extent, if any, of the problem. This may include the need for specialist examination and equipment or the opening up of parts of the building. All these activities may be beyond the scope of the agreed inspection, and it is reasonable for the surveyor to advise the client of these additional services and the cost that will be incurred if the further inspection is to be carried out.

The time-scale involved Any recommendation for action to be taken by a client must set out the time-scale for action and the consequence of inaction. In the case of a prospective purchaser, the surveyor will require the client to have the further inspection carried out before exchange of contracts. The consequence of failure to have that inspection is that the purchase may proceed at an inflated price and the client may have to pay for repairs in the immediate future.

In this way the client has a clear idea of the circumstances, the consequences and the actions available.

The duty of the surveyor has been interpreted by a number of judges, in court actions against surveyors, to include the duty to warn of a failure where there is evidence that a defect is common in buildings of a certain type, and where a defect is known to exist in similar properties in the same area.

The surveyor is faced on the one hand with avoiding an action for failing to point out the risk, while on the other trying to ensure that his report is objective advice not peppered with caveats and warnings.

To be certain that the client is aware both of the problem and of the surveyor's opinion of the extent of the risk, the surveyor should clearly state the risk that is present and then express an opinion as to the likelihood of the failure being present. For instance, if the building was in a terrace of houses, none of which showed any sign of failure, but where similar buildings across the road were known to have defective ties, the surveyor might report as follows:

We note that the external walls of the buildings are of cavity construction. We are aware that the property was constructed prior to 1981 and that there is a history of the failure of wall ties caused by corrosion in similar properties in this town. The failure of the metal ties will affect the structure of the building and make it vulnerable to progressive failure unless the ties are replaced.

The approximate cost of remedial treatment on this building will be in the region of £4,000. The cost of replacement of ties may vary, depending upon the ease or difficulty of carrying out the work. A builder's quotation should be obtained before exchange of contracts to determine the true cost of the works.

We have not inspected the ties in the wall and cannot advise you of their present condition.

During the course of our examination we did not see any indications of those failures that could indicate that the ties have already failed. We are aware that the failure of wire twist ties provides no advance warning. We are unable to state that the property is free from wall tie failure.

In the circumstances we recommend that the walls be opened up so that a random sample of ties may be inspected, to give some idea of the extent of deterioration that has taken place, before you exchange contracts for the purchase of this property. We will be pleased to carry out this work for you, for a fee of £200.

It is possible that the surveyor may feel that the risk of wall tie failure is remote. In that case, and if it is a fair judgement, based on sufficient experience and knowledge, then this advice should be passed on to the client. But the client should be made aware of the circumstances, and the financial risk that is involved if failure has occurred. The surveyor can then state, following the level of inspection that has been carried out, an opinion of the extent of the problem that he believes to be present in the particular case. If the surveyor is wrong, negligence will only be proved if it can be shown that the report and opinions stated in it were negligent. To have been negligent it will have to be shown that the surveyor:

failed to note defects that could have been seen during the course of the limited inspection that had been agreed between the parties;

failed to act upon information that was current at the time of the inspection;

failed to place a client on notice of the risk that was present, or could reasonably be expected to have been present, based upon information that was available from the inspection or that was published at the time.

By expressing the warning of what might be present, and advising on the further inspection that will identify the full extent of failure, the surveyor has disposed of the risk of a claim under any of the three areas set out above.

The inspection of a building, and the reporting on the condition of the building, is not an exact science. The careful structuring of a report will ensure that a client is aware of the risk that is present, or may occur in the future, and the actions that can be taken to help identify the extent of the problem.*

*A set of alternative phrases and paragraphs has been set out in Appendix 5.

APPENDIX 1   CHECK-LIST OF POINTS TO EXAMINE IN BUILDING SUFFERING WALL TIE FAILURE

The failure of wall ties can be caused by a number of events. The various types of tie will perform differently in each set of circumstances. It is therefore important that the type of tie has been identified. The examination of the building must take into account the risk that may be present as a result of any of the following.

Static Failures

Poor design   Are there any indications of design faults in the remainder of the building? Were the correct number of ties inserted?

Poor workmanship   Are there any signs of poor workmanship in the brickwork, such as an uneven wall face, perpends not in vertical alignment?

Chemical action   What are the components of the mortar and the brick? Are there any salt deposits?

Material failure   Are the ties to BS 1243?

Dynamic Failures

Alterations   Have any changes taken place, and if so what effect will they have had?

Weather   What is the exposure rating of the external walls, will it have changed, what tie was used, is it compatible with the risk caused by the exposure?
    Will high winds have weakened the wall, will the tie have been distorted, will there be a prospect of earlier failure?

Geological failures   Is there a risk of movement, or has there been movement?

Vibration   What vibration does the building suffer, has it increased, is it violent?

Impact   Has the building suffered any impact damage? If so, were any ties changed afterwards in the area affected?

APPENDIX 2 CONSIDERATIONS FOR THE SELECTION AND DESIGN OF CAVITY WALLS AND TIES

Ties used should comply with British Standard 1243, Part 2 of DD 140 or a British Board of Agrément certificate. This should ensure that the ties are durable. Stainless steel is recommended wherever there is any extra risk, and is preferable whenever possible.

Rigid ties should not be used if hard brickwork is being connected to lighter and softer blockwork. Damage will result from any movement. For large, more exposed structures, rigid ties are necessary, so the selection of materials for the cavity wall must take account of the imposed limitations and the requirement for matching strength.

The ties must be of the correct size. They should be long enough to give an embedment of at last 50 mm in each leaf on every occasion. Care has to be taken to ensure that the tie's dimension is adequate to cope with the variations that often occur in building work, even where this variation is in excess of normal acceptable construction standards.

The correct number of ties must be specified. The British Standards, BS 5628: Part 1: 1985 and BS 5628: Part 3: 1985, set out the standards. This amended standard now recommends that ties in cavity walls with one or both leaves of 90 mm or more should be placed at 900 mm horizontally and 450 mm vertically. With one or more leaves of 75 mm or less, the horizontal spacing is reduced to 450 mm. The number of ties per square metre is 2.5 and 4.9 respectively.

Ties should not be placed within 500 mm of the corners of a building. Extra ties must be put round openings and to the edge of gables. It is suggested that ties should be inserted at intervals of 300 mm vertically to the gable edge, as for vertical spacing around openings. Narrow panels of brickwork, e.g. less than a metre, may require ties every course.

Ties should be bedded into fresh mortar, not into part-set mortar beds. The tie must be kept clean and if there is any fall it must be to the outside, and not towards the inner leaf.

The ties should allow for sideways movement, caused by thermal or moisture variations, without causing damage to the materials of each leaf. It should have a drip to reduce the possibility of water crossing the cavity. The surface which is liable to catch mortar should be small, thus reducing the risk of the cavity being bridged. It must be thin enough to fit into a mortar bed, and yet give a good grip. These requirements have tended to lead to the use of the flexible butterfly tie or the double triangle.

Predicted life of zinc protection for galvanised mild steel wall ties

Standard of tie	years
Ties to current BS 1243 and DD 140 (1981 onwards)	47-93
Vertical-twist ties to BS 1243 before 1981	24-46
Wire ties to BS 1243 before 1981	13-26
Non-standard wire timber-frame ties	7-16

APPENDIX 3   CEMENT AND MORTAR
TESTING LABORATORIES

The following laboratories will carry out an examination of cement and mortar products to advise upon their chemical content:

Aberdeen Concrete Co Ltd.
Greenbank Road, Tullos
Aberdeen
Scotland AB1 4BQ
Tel: (0224) 871444

Albury Laboratories Ltd.
The Old Mill, Albury
Guildford
Surrey GU5 9AZ
Tel: (048 641) 2041

Bedfordshire County Council
Engineering Laboratory
Austin Canons Depot, Bedford Road
Kempston
Bedfordshire MK42 8AA
Tel: (0234) 45493

British Ceramic Research Association Ltd.
Queens Road, Penkhull
Stoke-on-Trent
Staffordshire ST4 7LQ
Tel: (0782) 45431

British Gypsum Ltd.
Research & Development Department
East Leake
Loughborough
Leicestershire LE2 6JQ
Tel: (0602) 214321

British Standards Institution
Test House, Marylands Avenue
Hemel Hempstead
Hertfordshire HP2 4SQ
Tel: (0442) 3111

The Chatfield Applied Research
    Laboratories Ltd.
13 Stafford Road
Croydon
Surrey CR0 4NG
Tel: (081) 688 5689

Fairclough Civil Engineering Ltd.
Geotechnical Services (Northern Div.)
Chapel Street
Adlington
Lancashire PR7 4JP
Tel: (0257) 480264

Fire Insurers' Research & Testing Organisation
Melrose Avenue
Borehamwood
Hertfordshire WD6 2BJ
Tel: (081) 207 2345

F.H.Gilman Ltd.
Bolton Hill, Tiers Cross
Haverfordwest
Dyfed, Wales SA6 3ER
Tel: (0437) 890481

Hampshire County Public Analyst &
    Scientific Adviser
Hyde Park Road
Southsea
Hampshire PO5 4LL
Tel: (0705) 828965

Hepworth Pipe Co Ltd.
Hazlehead, Stocksbridge
Sheffield
South Yorkshire S30 5HG
Tel: (0226) 763561

Ibstock Building Products Ltd.
Technical Services Department Laboratory
Ibstock
Leicestershire LE6 1HS
Tel: (0530) 60531

Industrial Science Division
Dept. of Economic Development
Antrim Road
Lisburn
County Antrim
Northern Ireland BT28 3AL
Tel: (084 62) 5161

Laing Design & Development Centre
Page Street
London NW7 2ER
Tel: (081) 959 3636

Sir Alfred McAlpine & Son (Southern) Ltd.
Kingswood Laboratory
Holyhead Road
Albrighton
Wolverhampton
Staffordshire WV7 3AR
Tel: 081-573 2637

Minton, Treharne & Davies Ltd.
Merton House, Bute Crescent
Cardiff
South Glamorgan, Wales CF1 6NB
Tel: (0222) 489002

Nicholas Colton & Partners
7-11 Harding Street
Leicester LE1 4BH
Tel: (0533) 536333

Harry Stanger Ltd.
The Laboratories, Fortune Lane
Elstree
Hertfordshire WD6 3HQ
Tel: (081) 207 3191

Taylor Woodrow Construction Ltd.
Research Laboratories
Taywood Engineering Ltd.
345 Ruislip Road
Southall
Middlesex UD1 2QX
Tel: (081) 575 4509/4849

Timber Research & Development
    Association
Stocking Lane, Hughenden Valley
High Wycombe
Bucks HP14 4ND
Tel: (024 024) 3091

Warrington Research Centre
Holmesfield Road
Warrington
Cheshire WA1 2DS
Tel: (0925) 55116

Wimpey Laboratories Ltd.
Metallurgy Laboratory and Structures
    Laboratory
Beaconsfield Road
Hayes
Middlesex UB4 0LS
Tel: (081) 573 7744

Yarsley Technical Centre Ltd.
Trowers Way
Redhill
Surrey RH1 2JN
Tel: (0737) 65070/9

APPENDIX 4 SUPPLIERS OF REPLACEMENT TIES

Copper helical bar
Talbot Helifix Ltd.
Lennox House
3 Pierrepont Street
Bath BA1 1LB
Tel: (0225) 462660

Grouted sleeved stainless steel &
Grouted non-sleeved stainless steel
Pynford Chemical Applications Ltd.
Warlies Park House
Upshire, Waltham Abbey
Essex EN9 3SL
Tel: (0992) 764213

WT Fixings (Holdraven Ltd.)
4 Marina Centre
Brighton Marina
Brighton BN2 5UR
Tel: (0273) 609660

Stainless steel screw-in anchors
Hilti (GB) Ltd.
Hilti House
Chester Road
Manchester M16 0GW
Tel: (061) 872 5010

Red Head
Queenslie Industrial Estate
Glasgow G33 4BL
Tel: (041) 774 2267

Harris & Edgar Ltd.
Delta Progress Works
222 Purley Way
Croydon CR9 4JH
Tel: (081) 686 4891

Combined bolt and resin ties
Suppliers as for Stainless steel screw-in anchors

Stainless steel ties with plastic friction grips on both ends
Suppliers as for Stainless steel screw-in anchors

Stainless steel grouted
Hilti (GB) Ltd.
Hilti House
Chester Road
Manchester M16 0GW
Tel: (061) 872 5010

APPENDIX 5   PHRASES FOR REPORTING UPON THE PRESENCE OF WALL TIE FAILURE

'During the course of our examination we noted a number of failures that suggest that wall tie corrosion is taking place. The failure of these metal ties reduces the stability of the walls of the building and can lead to failure of a part or all of the external walls of the building.'

'We note that the external walls of the building are of cavity construction. We are aware that the property was constructed prior to 1981 and that there is a history of the failure of wall ties in buildings erected between 1955 and 1981.'

'The failure of the metal ties will affect the structure of the building and make it vulnerable to progressive failure unless the ties are replaced.'

'The approximate cost of remedial treatment on this building will be in the region of £0,000. The cost of the replacement of ties may vary, depending upon the ease or difficulty of carrying out the work. A builder's quotation should be obtained before you exchange contracts so that the likely cost of replacing the ties can be determined.

We will be pleased to arrange for a contractor's quotation for the cost of the replacement of wall ties.
or

A list of firms who carry out the replacement of wall ties is attached to this report. We suggest that you contact one of the firms to obtain a quotation as to the cost of removing the existing ties and the installation of replacement ties.'

'We have not inspected the ties in the wall and cannot advise you of their present condition.'

'During the course of our examination we did not see any indications of failures that could indicate that the ties have already failed. Because we are aware that the failure of wire twist ties provide no advance warning, we are unable to state that the property is free from wall tie failure.'

'In these circumstances we recommend that the walls be opened up so that a random sample of ties may be inspected to give some idea of the extent of deterioration that has taken place before you exchange contracts for the purchase of this property.

We will be pleased to carry out this work for you, for a fee of £000.

or

We can arrange for this inspection to be carried out by (name of contractor) and wish to advise you that their charge for this work will be £000.'

INDEX